551.57
Br

W9-CFS-983

What Makes It Rain?
The Story of a Raindrop

Written by Keith Brandt
Illustrated by Yoshi Miyake

DISCARDED

Troll Associates

Library of Congress Cataloging in Publication Data

Brandt, Keith.
    What makes it rain?

    Summary: Follows the journey of a raindrop
through the water cycle and briefly discusses the
characteristics and importance of water.
    1. Rain and rainfall—Juvenile literature.
[1. Rain and rainfall. 2. Water. 3. Hydrologic
cycle] I. Miyake, Yoshi. II. Title.
QC9247.B68   551.57'81   81-7495
ISBN 0-89375-582-6   AACR2
ISBN 0-89375-583-4 (pbk.)

Copyright © 1982 by Troll Associates, Mahwah, New Jersey.
All rights reserved. No part of this book may be used or
reproduced in any manner whatsoever without written
permission from the publisher.
Printed in the United States of America.

10   9   8   7   6   5   4   3   2   1

Water is everywhere. It falls to earth as rain. It makes puddles, streams, rivers, lakes, and oceans. Water runs down from the tops of mountains. It flows under the ground. It is in every living thing.

When you see a raindrop, do you ever wonder where it comes from? It comes from the sky, where it is carried in a cloud. Like ships on the ocean, clouds move through the sky. They are blown by the wind. Clouds travel for miles and miles. When a cloud grows heavy and full of drops of water, it rains.

Sometimes, the rain comes down in big, fat drops. Sometimes, it comes down in very small droplets.

Here comes a drop of rain falling through the sky. Down, down, down it falls. At last, with a small splash, it hits the ground. Soon it is joined by thousands of other raindrops. Together they make a puddle.

The rainstorm ends, and the sun comes out. The sun's hot rays reach the puddle of water. They warm the water, just the way water is heated on a kitchen stove. After a while, the puddle dries. The water seems to have disappeared. It has gone back into the air, where it came from. How did this happen?

The heat of the sun changed the water into a gas. Most of the time you cannot see water when it is a gas. This gas is called *water vapor*. The change is called *evaporation*.

What happens after the raindrop in the puddle is changed to a gas? It rises in the air. The wind takes it, and other raindrops, back into the sky. After a while, it will be part of a cloud again.

When the wind is strong, the clouds seem to race through the sky. One drop of rain is carried far away by the wind. The raindrop can travel all the way across the country, over the ocean, even to another country. But sooner or later, it will fall to earth again.

If the air around the cloud is very cold, a drop of water will fall to earth as a snowflake. Snow and ice are what we call water when it is a *solid*. Just as heat turns water into a gas, cold turns water into a solid.

If the ground is very cold where the snowflakes fall, they will not melt. Soon the ground will be covered by a white blanket of snow.

If the ground is not cold, the snow will melt. The snow will also melt if the sun comes out and warms the air. Some of the snow will change back into a gas, and rise into the air. Some of the snow will become water, which sinks into the ground. The water that goes into the ground will stay there all winter. In the spring, the growing plants will use this water.

Plants cannot live without water. Animals cannot live without water. People cannot live without water. Water, like the air we breathe, is very important to all living things.

Water is like air in another way. It is all around us, all the time. When water is in the air, you usually cannot see it. But it is there. Most of the time you *can* see water. When you look at a lake, you are seeing many, many drops of water. Many drops of water all together are called a *liquid*.

When water is a liquid, you can drink it, pour it, swim in it, or play in it. For water to stay a liquid, it must not be too hot or too cold.

Heat and cold make water change its form. But no matter what form it is in—liquid, solid, or gas—water is always water.

And it doesn't only change form; water also travels everywhere. Sometimes it moves from one place to another in liquid form—as a raindrop, in a stream, or in the ocean.

Sometimes water moves in solid form—as an iceberg in the ocean, as a snowflake through the sky, even as a snowball thrown at a tree.

And sometimes water moves in the form of a gas—as a cloud, as a mist rising over a lake, or as steam puffing from the spout of a teakettle.

All the moves and all the changes a drop of water goes through are part of something called the *water cycle*. The word *cycle* means "circle."

The water cycle is a journey every drop of water makes. It is a journey that can take one raindrop all the way around the world and bring it right back to where it started.

DISCARDED

The raindrop you watch fall from the sky today may have been in a faraway ocean last year. It has gone through many changes and traveled many miles—all part of the water cycle.

The water cycle is very important. Without water, plants cannot grow. None of the crops we eat would grow if the water cycle stopped—they must have rain.

Some plants need very little water. A cactus is one kind of plant that needs only a very small amount of water. That is because it can store water for a long time. A cactus grows best in places that are very dry, like the desert, where almost no other plants can grow.

Animals cannot live without water either. They drink it from streams. They also get it in their food; just as we do. Animals are part of the water cycle, too. After they take in water, they give it back in many different forms. Cows give back a lot of it as milk. Hens give back a lot of it in eggs.

Some animals have special ways of living in places where there is not much water. Camels can live very well in very dry places. They can go without water for days. This makes the camel a perfect desert animal.

Plants get water from the ground and the sky. People and animals get water in many ways. You get water from the food you eat. You also get water by turning on the faucet in your kitchen. But whether it comes from food or a faucet, you need water to grow and be healthy. Water is an important part of you.

How much do you weigh? Well, you may be surprised to learn that two-thirds of your weight is water. There is water in every part of your body.

When you turn on the cold-water faucet, water pours out. But how did it get there?

The water you drink may have started as snow falling on a mountain far away from your home. It had to take a long journey before it came to you.

Every winter, snow falls on that faraway mountain. In the spring, the snow melts. It runs down the mountain in little streams. These streams run into bigger streams. And most of the water in these bigger streams runs into rivers. But not all of it. Some of the water flows into a large, special kind of lake called a reservoir.

The water in the reservoir flows through pipes that are buried underground. These pipes go to your town. One of those pipes goes to your house. That is how, when you turn on the faucet in the kitchen, water from the reservoir comes to you.

The reservoir almost always has a big supply of water. As you are using it, it is being replaced by rain and snow in the mountains far away. But sometimes there is not much rain and snow. When the water supply is low, it is called a drought. When there is a very bad drought, plants do not grow, and there is less food for us and all other animals.

There are also times when too much rain and snow fall. When there is more water than the rivers and the ground can hold, we have floods.

Floods harm many plants. Flood waters also damage homes and highways. So water, which does so many good things for us, can also do damage. Luckily, most of the things it does are good.

Does all water taste the same? No. Water from different places tastes different. It is not the water itself, but what is in the water that gives it a taste.

Most of the water in the world has a salty taste. That is because most of the water in the world is in the oceans. Much of the water that flows in rivers is carried to the oceans.

The water that comes from the rivers to the oceans is not salty. But it becomes salty as soon as it reaches the ocean. This is because there are huge amounts of salt in the ocean. When the sun warms the surface of the ocean, some of the water evaporates into the air. The salt does not evaporate with the water. It stays in the ocean. If all the waters in the oceans evaporated, mountains of salt would be left behind.

Experiment:
Add some salt to a glass of water.

1

2
Water evaporates.

3
The salt stays.

We do not have to worry that the oceans will disappear. As the sun evaporates some ocean water, clouds are bringing rain to the ocean. Rivers are bringing water to the ocean. The ocean water cycle never stops.

The next time you see a raindrop, think about the special trip it's been on. Up and down in the air and all around the world in an endless journey—each raindrop is an important part of life on earth.

DISCARDED